The Origin, Evolution and Future of the Sun

Contents

Chapter I   Introduction
1. Overview
2. The Basic Facts of the Sun
3. Chemical Composition of the Sun
4. The Rotation of the Sun
5. The Revolving of the Sun
6. The Stars Most Close to the Sun

Chapter II   Structure of the Sun
1. Photosphere
2. Chromosphere
3. Corona

Chapter III   The Solar Activities
1. Sunspot
2. Solar Flare
3. Solar Facula
4. Granules
5. Solar Wind
6. Coronal Holes
7. Sunlight

Chapter IV   The Origin and Evolution of the Sun
Chapter V   The Future of the Sun
Chapter VI   The Possible New Sun

Chapter I   Introduction

1. Overview

The Sun is ordinary as it is only a medium star in the universe; the Sun is great as it nurtures the human and other living things on the Earth. All things, including the Sun, will experience the process from birth and growth to decline and end.

The Sun is the mother star and central celestial body of the Solar System. The nine planets, asteroids, meteors, comet, Trans-Neptunian objects and Cosmic Dusts, etc. revolve around the Sun under the powerful gravity of the Sun, while the Sun revolves around the center of the Galaxy or the galactic center. Among more than 100 billion stars in the Galaxy with a diameter of about 100,000 light years, the Sun is just a very ordinary one. It is located on the Orion Arm in the north of galactic plane and is about 30000 light years away from the galactic center and about 26 light years away from the north of galactic plane. The Sun revolves around the galactic center at the speed of 250km/second, the period of revolving one cycle is about 250 million years, meanwhile it moves towards the vicinity of the Vega at the speed of 19.7km/second. The Sun also rotates, the period of rotating one cycle is about 25 days in the heliographic equatorial zone and about 35 days in the two polar zones.

The Sun is a huge and glowing gaseous star with continuous thermonuclear reaction inside, its surface temperature is about 6000k and the temperature in the core of the Sun reaches 15 million K. The Sun is an idea globe intertwined with plasma and magnetic field. It is the star most close to the Earth in the Galaxy and its apparent magnitude reaches -26.8, becoming the brightest celestial body visible on the Earth. The angular diameter of the solar disk in the sky is 32 minute of arc, it is very close to the Moon's angular diameter visible from the Earth. It is a fantastic coincidence, which makes the solar eclipse extra spectacular.

The orbit of the Earth revolving around the Sun is elliptic, the Earth is farthest away from the Sun in July (known as Aphelion) and most close to the Sun in January (known as Perihelion), the average distance is 149.60 million kilometers (the distance is called One Astronomic Unit in in astronomy, AU for short). Calculating by the average distance, it takes the sunlight 8 minutes and 19 seconds to reach the Earth. The energy contained in the sunlight supports the growth of all the living things on the Earth and also dominates the climate, weather and natural environment of the Earth.

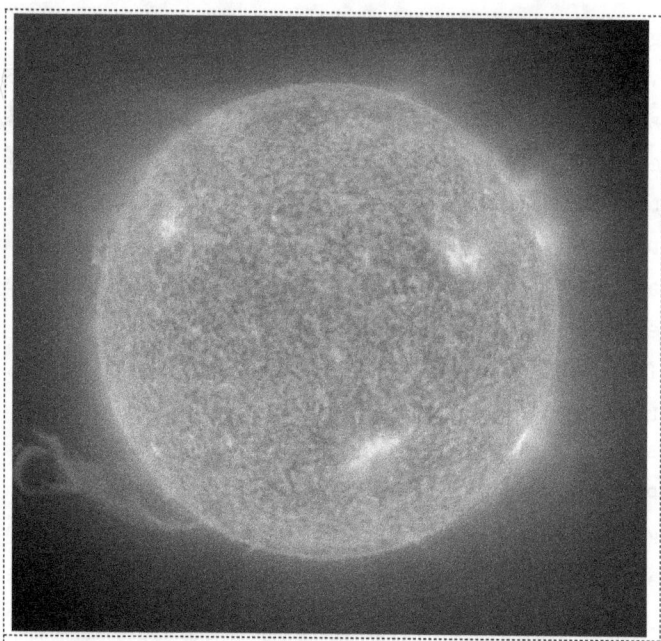
The Sun

The Solar System

2. The Basic Facts of the Sun

In the universe known by so far, the Sun is just a medium-sized glowing and gaseous star although it is the mother star of the solar system and is the only one celestial body that can shine in the solar system. After knowing the distance between the Sun and Earth and measuring the visual angle diameter of the solar disk on the Earth, with use of simple triangle relation, it can be worked out that the radius of the Sun is 696,000 km, the diameter is 1392000 (1.392×10^6) km, equal to 109 times the Earth diameter, hereby it can be worked out that the volume of the Sun is about 1.30 million times the same of the Earth.

The astronomers, by referring to the third law of Kepler's planet motion, the mass of the Earth, the radius of the Earth orbit revolving around the Sun and the period of revolving one cycle, worked out that the mass of the Sun is 1.989×10^{30} kg (1.989×10^{27} metric tons), equal to 33,000 times the mass of the Earth and accounting for 99.86% of the mass of the whole Solar System. The surface area of the Sun is about 12,000 times the same of the Earth. Although the Sun is a huge celestial body compared with the Earth, it is no more than a very ordinary star with medium mass in the boundless universe.

Based on the volume and mass of the Sun, it can be worked out that the average density of the Sun is 1.409 g/cm³, about 0.26 times the average density of the Earth. The density of the outer visible portion of the Sun is about one millionth of the water density; the density of the central portion of the Sun is 85 times bigger than the water density.

The gravity acceleration of the Sun surface is 273.9810 meters/Second2 and it is about 28 times the same of the Earth surface. If a person stands on the surface of the Sun, his weight will be 20 times heavier than the weight on the Earth. The escape velocity of the Sun surface is about 617.7km/second, it means that no neutral particle can escape from the gravity of the Sun and fly into the outer space unless it moves at a speed exceeding this value.

Summary of the Parameters of the Sun

Indicators	Parameters	Indicators	Parameters
Class	Star	Mass	1.9891×10^{30} kg (333,400 times the mass of the Earth, accounting for 99.86% of the mass of whole solar system)
Average Density	1.408×10^3 kg/m³ or 1.409 g/cm³, about one fourth of the density of the Earth	Diameter	1.392×10^6 km (About 1.40 million km, 109 times the diameter of the Earth)
Density of Outer Visible Portion	About 1.3g/cm³	Core Density	About 160g/cm³, 85 times bigger than the density of water
Density Relative to the Earth	Approximately 0.26	Density Relative to Water	Approximately 1.3
Surface Temperature	About 6000 °C	Radius	696,000km (Approximately 110 times the Radius of the Earth)

The Origin, Evolution and Future of the Sun

Surface Area	Approximately 6.09 × 10^{12} Square Kilometers (About 12,000 times the surface area of the Earth	Volume	Approximately 1.412 ×10^{18} cubic kilometers (1,300,000 times the Volume of the Earth)
Spectral Class	G2V	Surface Gravity Acceleration	2.74×10^2 m/second^2 (27.9 times the Surface Gravity Acceleration of the Earth)
Escape Velocity	617.7 km/s	Right Ascension	286.13°
Rotation Period	25.05 Days	Solar Declination	+63.87°
Orbital Period	(2.25-2.50) ×10^8 a	Bolometric Magnitude	-26.82
Mean Earth-Sun Distance	(1 Astronomical Unit, AU): 1.49597870×10^{11} meters (150 million km)		
Maximum Distance between the Sun and the Earth	1.5210×10^{11} meters	Apparent Magnitude	(V) -26.74

The Origin, Evolution and Future of the Sun

Minimum Distance between the Sun and the Earth	1.4710×10^{11} meters	Absolute Magnitude	4.83
Distance Difference between Aphelion and Perihelion			5 million km
Core Temperature	Approximately 15 million ℃	Life Span of the Sun	About 10 Billion Years
Age	About 4.6 Billion Years	Coronal Temperature	About 1,000,000 ℃
Luminosity (LS)	About 3.827×10^{26} Js^{-1}	Solar Activity Cycle	11.04 Years
Total Radiated Power	3.86×10^{26} W	Effective Temperature	5800 ℃
Sun Movement Speed (Direction α=18h07m, δ=+30°) = 19.7km/second		Total Radiation Power: 3.86×10^26 W(Joule/Second)	

The peak wavelength (500 nanometers) of solar radiation stands at the transition area between the blue and green in the light spectrum. The temperature of a star is closely associated with the dominant wavelength in its radiation. The surface temperature of the Sun is about 6000k. As the eyes of human are more sensitive to other colors around the peak wavelength, the Sun looks yellow or red in the eyes of human.

3. Chemical Composition of the Sun

With respect to its chemical composition, the Sun comprises hydrogen (71%), helium (26%) and a few heavy elements. They release the energy (light and heat) through nuclear fusion. According to the relevant theory, the substances produced by the nuclear fusion are the metals, including iron and copper, etc.

The Chemical Composition of the Sun

Elements	Proportions	Elements	Proportions
Hydrogen	73.46%	Neon	0.12%
Helium	24.85%	Nitrogen	0.09%
Oxygen	0.77%	Silicon	0.07%
Carbon	0.29%	Magnesium	0.05%
Iron	0.16%	Sulphur	0.04%

4. The Rotation of the Sun

Like other celestial body, the Sun also rotates around its own axis from west to east. However, the observation and research findings indicate that the rotation speed of the Sun is not same at different latitudes. At the equator, it takes the Sun 25.4 day to complete one rotation; at the latitude 40, it takes the Sun 27.2 days to complete one rotation; at the polar zones, it takes the Sun 35 days or so to rotate one cycle. Such rotation means is called "Differential Rotation".

The reason why the Sun rotates is that the whole solar system is outcome of the shrinkage of a rotating nebula. As the nebula rotates itself, the Sun transformed from the nebula rotates naturally. As the volume of the Sun is far smaller than the nebula, the Sun rotates much faster than the nebula. The reason why the nebula can rotate is that there is turbulent eddy current in the nebula, propelling the rotation of the nebula.

5. The Revolving of the Sun

The Sun revolves around the Galactic Center and the revolving period is about 2.5×10^8 years. There may be a huge black hole in the Galactic Center. As numerous stars overspread around the center of the Galactic System, the Galactic Center looks like a "Galactic Disk". These stars revolve around the "Galactic Nucleus". Different from the Earth revolving around the Sun, these stars will move closer to the "Galactic Nucleus" when they complete one revolving.

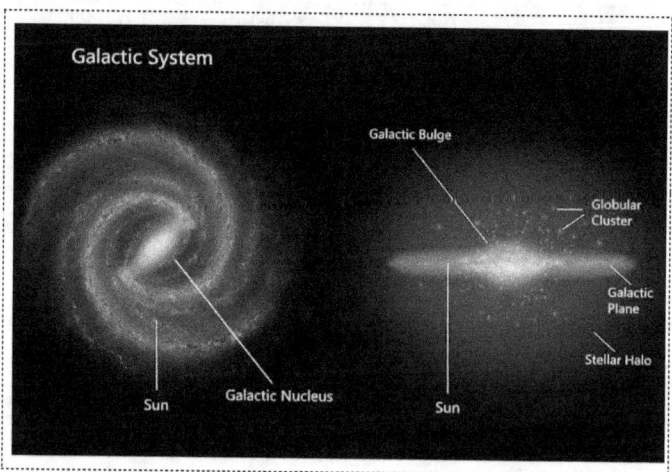

The Position of the sun in the Galactic System

The Sun is the star most close to the Earth. It is located nearby the symmetry plane of the Galactic System and about 26000 light years away from the center of the Galactic System. In the position of about 26 light years away from the galactic plane, the Sun revolves around the center of the Galactic Center at the speed of 250km/second and move towards the vicinity of the Vega at the speed of 19.7km/second.

The Sun is passing through the Local Interstellar Cloud in the local bubble zone of inner-edge Orion Arm of the Galactic System. Within the distance of 17 light years away from the Earth, there are 50 stars adjacent to the Earth, a star most close to the Earth is the red dwarf star, and it is called "Proxima Centauri", about 4.24 light years away from the Earth. The Sun ranks the 4th position among these stars in mass. The Sun revolves around the Galactic System at the position of 24000 – 26000 light years away from the Galactic Center or "Galactic Nucleus". Looking at the Sun from the Galactic North, the Sun moves on a clockwise orbit and revolves one cycle about 225 to 250 million years.

As the Galactic System moves towards Hydra Hya at the speed of 550km/second in the cosmic microwave background (CMB), after these two speeds are integrated, the Sun moves towards the Crater or Leo at the speed of 370km/second relative to the CMB.

6. The Stars Most Close to the Sun

The Alpha Centauri is a stellar system (a triple star system) most close to the Sun, it is only 4.24 light years away from the Sun (about 277600 AU), the Proxima Centauri is usually considered as the member of the stellar system and it is only 4.22 light years away from the Sun, the star most close to the Sun. If someone stands on the Proxima Centauri, he can see three suns.

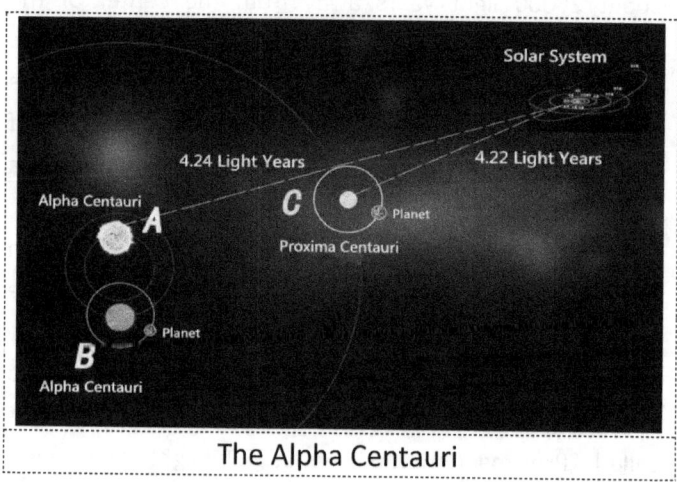

The Alpha Centauri

Watching the Sun from the position of the Alpha Centauri, the Sun becomes a star with apparent magnitude of 0.5 in the Cassiopeia. Generally speaking, the shape of Cassiopeia turns from \/\/ into /\/\/, the Sun is located at the end of Segin.

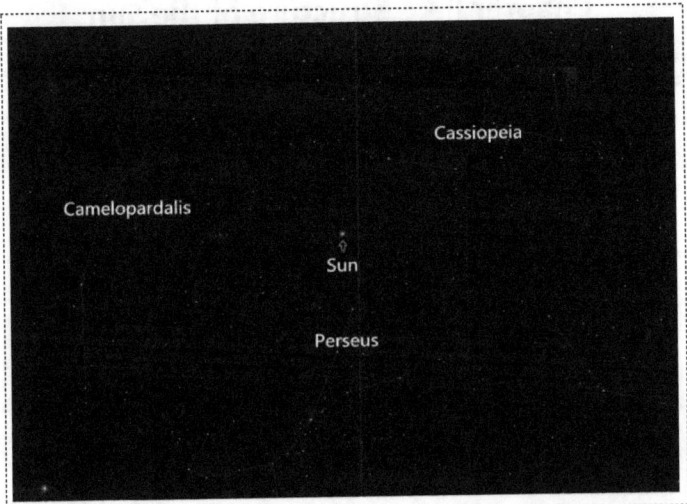
Watch the Sun from the Position of the Alpha Centauri

Chapter II   Structure of the Sun

From center away, the Sun comprises the core (Nuclear Reaction Zone), radiative zone, convective zone, photosphere, chromosphere and corona in structure. Below the photosphere is called "Solar Interior", namely the core, radiative zone, convective zone from inside out; above the photosphere is called "Solar Atmosphere", namely photosphere, chromosphere and corona from inside out. The Solar Atmosphere is directly visible to the human.

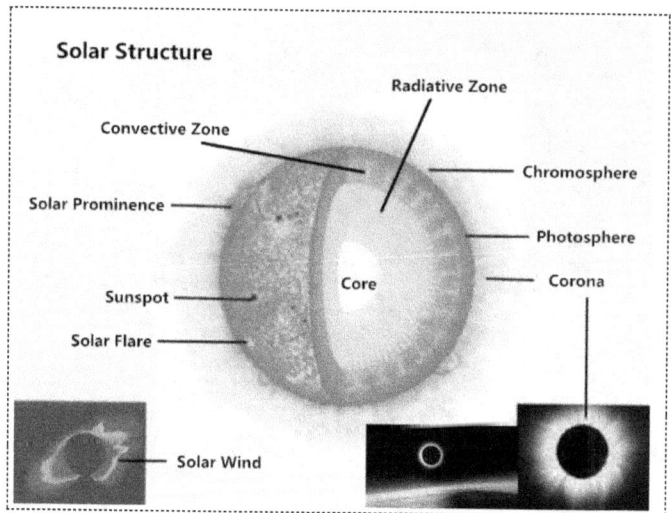

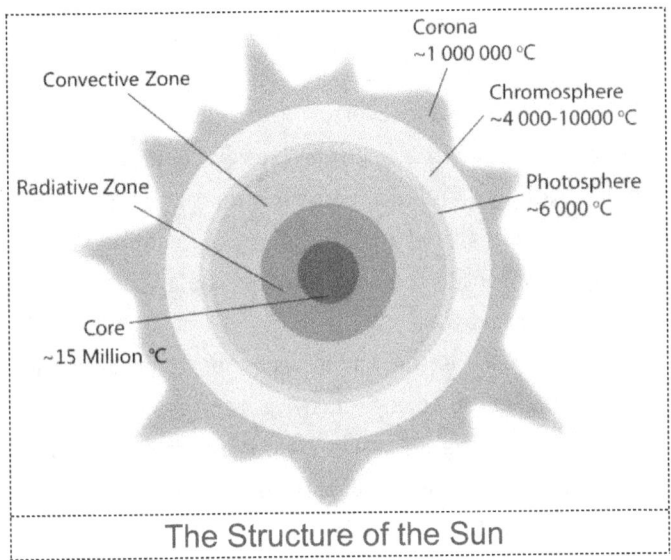

The Structure of the Sun

The "Solar Interior" is divided into three zones: Core, Radiative Zone and Convective Zone.

The radius of the Core is one fourth of the radius of the sun, the Core accounts for more than half of the mass of the whole Sun. The temperature at the Core is as high as 15 million °C, the pressure is equal to 300 billion barometric pressures, the super-high temperature and pressure trigger the thermonuclear reaction of four hydrogen nuclei fusion into one helium nucleus all the time, thus releasing the huge energy. The energy transmits outward through the matters in the radiation zone and convection zone, reaches the bottom of the Photosphere and radiate outward through the Photosphere. The matter density in the Core is as high as 160g per cubic centimeter. Under the attraction by its own powerful gravity, the Core of the Sun stays in a high-density, high-temperature and high-pressure state, it is the source of the huge energy transmitted by the Sun and is also called "Nuclear Reaction

Zone". Pursuant to the nuclear physics and mass-energy conversion formula E=mc², in the core, 600 million tons of hydrogen is converted into 596 million tons of helium per second, releasing the energy equal to 4 million tons of hydrogen and giving out the heat and light to the planets, including the Earth. Based on the estimate of the hydrogen content in the Sun, the Sun has a life span of 5 billion years at least.

Next to the Core is the Radiative Zone, its range stands between 0.25~0.86 solar radius. The temperature, density and pressure in the Radiative Zone diminish from inward to outward. In respect of volume, the Radiative Zone accounts for vast majority of the volume of the whole Sun. The Radiation Zone contains various electromagnetic radiations and particles flow. The transmission process of radiation is a process in which the energy is absorbed by the matters and retransmitted many times. In the way from the Nuclear Reaction Zone to the surface of the Sun, the energy radiates outward in the forms of X-ray, far-ultraviolet ray, ultraviolet rays and visible light in turn. The Sun is the inexhaustible energy source.

Outside the Radiation Zone is the Convective Zone, it is most outward zone of Solar Interior and its range stands between 0.71 solar radius and the bottom of Solar Atmosphere or photosphere. The depth of the Convective Zone exceeds 100 thousand meters. As the temperature, pressure and density gradient in the Convective Zone are very great, the properties of solar gases vary greatly and are very unstable, the radial convective motion of the matters is very strong, the hot matters move outward and cold matters sink into the interior, the solar gases form the obvious up-down flow. The energy produced in the Core of the Sun is transmitted outward through such convection flow, besides the outward radiation.

The parts above the photosphere are called "Solar Atmosphere" collectively. The Solar Atmosphere, like the atmosphere of the Earth, can be divided into several layers based on different heights and properties, namely the photosphere, chromosphere and corona from inside out. The solar surface visible to the human is the bottom of the Solar Atmosphere, its temperature is about 6000 ℃. As it is not transparent, human cannot see the internal structure of the Sun directly. The Solar Atmosphere transcends the whole electromagnetic spectrum, ranging from radio, visible light to gamma rays.

1. Photosphere

The solar atmosphere above the Convective Zone is called solar photosphere, it is the solar disk seen by human. The solar radius in official definition refers to the radius of the photosphere. The photosphere is located outside the Convective Zone and it is the bottom or most inner layer of the solar atmosphere. The surface of the photosphere is gaseous and its mean density is only one several hundred millionth of the water density. As its depth is as long as 500km, the photosphere is non-transparent, it is a thin gaseous layer. It defines the very clear boundary of the Sun, almost all the visible lights are transmitted from the photosphere.

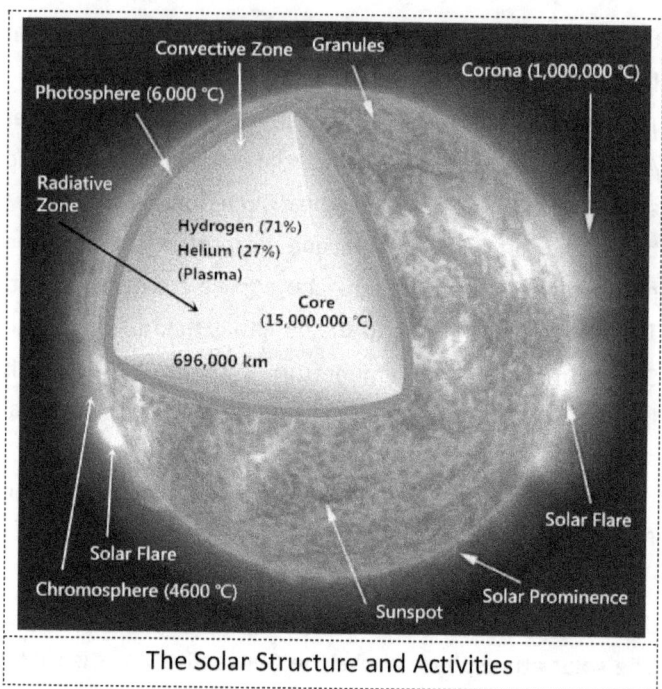

The Solar Structure and Activities

2. Chromosphere

The chromosphere is above the photosphere. It is about 8,000 km in depth and is as same as the photosphere in chemical composition, but matter density and pressure in the chromosphere are far much lower than the photosphere in this respect. The solar temperature declines gradually from the core to the photosphere, temperature of the boundary between the photosphere and chromosphere is about 4300°C. However, in the chromosphere, the temperature rises greatly, the temperature of the chromosphere top is as high as tens of thousands Celsius degrees and the temperature of the Corona even reaches one million Celsius degrees. Such phenomenon contrary to the common sense remains a puzzle today and no right answer is found.

As the visible light transmitted by the chromosphere is less than 1% of the visible light transmitted by the photosphere, the chromosphere cannot be seen by the human until the total solar eclipse comes, namely in a short moment lasting only several seconds when the Moon totally covers up the bright photosphere, a narrow and rose red light ring appears on the edge of solar disk, it is the chromosphere. At other times, the chromosphere can be watched with aid of monochromatic chromospheric telescope (the wave length is 6563 angstrom).

The Chromosphere

Besides, human can see many rising flames on the chromosphere, they are called "Solar Prominence" astronomically. The solar Prominence is a gas current erupted from the solar surface and it is a very strong and fast-changing solar activity occurring on the chromosphere. A complete solar prominence lasts tens of minutes. The solar prominence is various is shape, some are like floating clouds and smokes, some are like waterfall and fountain and some are like arch bridge and tussock, etc.

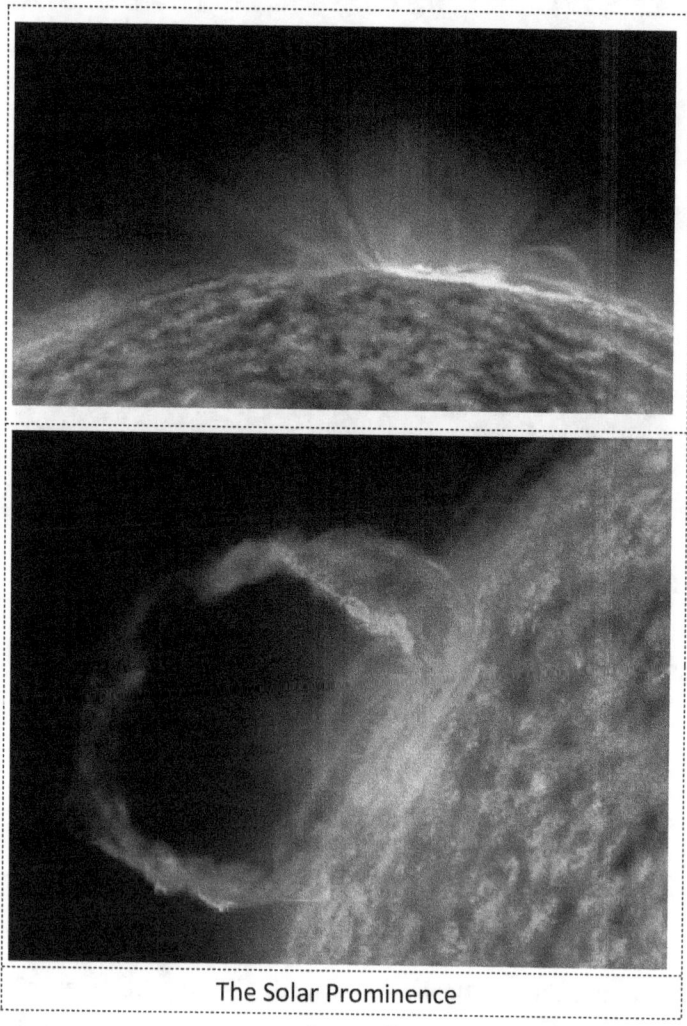

The Solar Prominence

3. Corona

The Corona is above the chromosphere, it is the most outward zone of the solar atmosphere and its range can stretch to the length several times the solar radius. The Corona is made up of high-temperature and low-density plasma. The luminosity of the Corona is very weak and it is lower than 1% of the luminosity of the solar disk, equal to the brightness of a full moon. So, only when the total solar eclipse comes, it is visible to the human: i the dark sky, the Moon covers up the solar disk, a bluish white light zone appears around the solar disk, it is the Corona – the most outward solar atmosphere.

The Corona

At other times, the Corona can be seen with aid of special coronagraph. The temperature of the Corona is as high as one million Celsius degrees, the shape and size of the Celsius degrees are associated with solar activities. In the "Solar Maximum Year", the shape of the Corona is close to a circle; in the "Solar Minimum Year", it presents an ellipse.

As the matters in the Corona are very thin, the Corona expands outward all the time, the hot ionized gas particles flows outward continuously, thus forming the solar wind.

Chapter III The Solar Activities

The intensive activities are going on all the time inside the Sun although it seems very quiet. In the core, namely the Nuclear Reaction Zone, thermonuclear reaction produces the light and heat radiated to the outer space, becoming the primary energy source of the Earth. The solar activities are principally going on in the Solar Atmosphere and they include Sunspot, Solar Flare, Solar Facula, Granules, Solar Wind and Coronal Holes, the solar activities produce great impacts on the Earth and human being.

1. Sunspot

The sunspot is the area that has relatively low temperature and appears to be black on the solar surface. The human ancestors 4000 year ago watched the black spots like crow with three legs on the Sun, these often-seen black spots are called "Sunspot" by modern people. Sunspot is one important activity of the photosphere. Sunspot is the huge wind vortex on the photosphere, it is the local strong magnetic field formed by the intense activities of the photosphere. Most of the sunspots present oval shape. They appear to be black relatively under the bright photosphere background, but their temperature is as high as 4000 ℃. The energy of one sunspot is equal to the light of a full moon. The sunspots on the photosphere are under constant changes in size, number, position and shapes, etc., which reflect the changes of the solar radiation. The sunspot change has complex periodic phenomenon, the average solar cycle is 11.2 years, it is also the solar cycle of the whole Sun. The year in which the sunspots are most called "Solar Maximum Year", the year in which the sunspots are least is called "Solar Minimum Year".

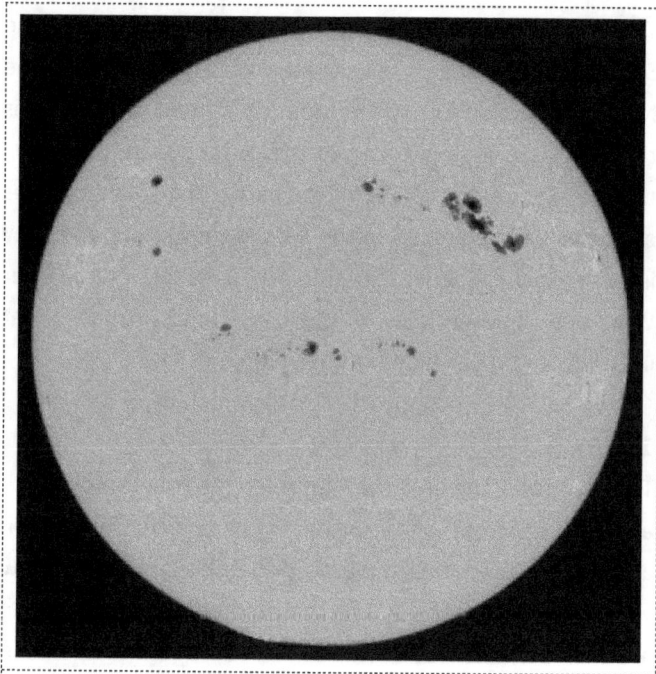

The Sunspots on the Photosphere

The sunspot produces great impacts on the Earth and human health:

(i) It interferes the magnetic field and ionosphere of the Earth, causing the malfunction of the compass, straying of animals, severe damage or interruption of radio communication, and directly endangering the communication system security of the aircraft, ships and man-made satellite;

(ii) It harms the human health. During a period of 803

years from AD 1173~1976, the epidemic influenza occurred 56 times in the Solar Maximum Years. Furthermore, the number of people who died of myocardial infarction (heart attack) increased dramatically in Solar Maximum Year;

(iii) In Solar Maximum Years, the Sun will transmit massive high-energy particles and X-rays, triggering the geomagnetic storm and leading to the Earth climate anomalies, the microorganisms are produced on mass scale, providing hotbed for the outbreak of pandemic.

(iv) Sunspot activity can trigger the ionization of organism substance and cause the genetic variation or mutation inheritance in Influenza virus, producing subtype of influenza virus with strong infectivity, leading to the outbreak of epidemic influenza or other complex biochemical reactions.

2. Solar Flare

The Solar Flare is a drastic solar activity and it is a process in which the solar energy is released in high concentration. It is generally believed that the Solar Flare occurs in the chromosphere, so it is also called "chromospheric eruption". Its main features are: on the solar disk (usually over the sunspots), a fast-developing bright spot shining comes out suddenly, its luminance rises quickly and falls slowly, lasting a period of several minutes to tens of minutes. In Solar Maximum Year, the Solar Flare occurs more frequently and intensely.

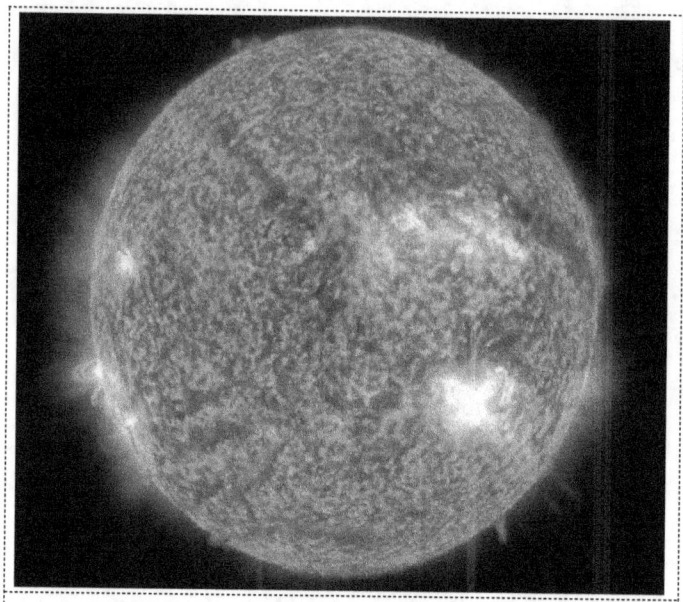

The Solar Flare

The energy released by a Solar Flare amazing, it is equal to the total energy released by 100,000 to one million strong volcanic eruptions, or equal to the energy released by the explosion of 10-billion 100-ton hydrogen bombs. A strong Solar Flare can release the energy of 10^{25} Joules.

Besides the sudden brightening, the Solar Flare contributes to the sudden intensification of radiation flux from radio wavelength band to X-ray. The radiations transmitted by the Solar Flare are various in variety, besides the visible lights, they include the ultraviolet light, X-ray and gamma ray, infrared ray and radio radiation. The Solar Flare also produces shock waves, high-energy particles and even the cosmic radiation with extra high energy.

The Solar Flare produces great impact on the Geospace Environment. When the Solar Flare erupts, massive high-energy particles move close to the Earth orbit, the safety of the instruments and astronauts in the space crafts will be severely endangered; when the Solar Flare radiation comes near to the Earth, it collides with the atmospheric molecules violently, damaging the ionosphere and paralyzing ionosphere's function of reflecting the radio wave. The radio communication, particularly the short-wave communication, and TV station and radio broadcasting will be interfered and even be paralyzed. Under the coaction of high-energy charged particles stream transmitted by the Solar Energy and high-layer earth atmosphere, the aurora is produced, it interferes the geomagnetic field, triggering the magnetic storm.

Besides, the Solar Flare also produces the direct or indirect impact of different levels on the weather and hydrography. For this reason, the scientists are working hard for unveiling the secret of the Solar Flare.

3. Solar Facula

The Solar Facula is a solar activity on the edge of the photosphere, it refers to the bright spot appearing on the edge of the photosphere. With aid of telescope, it can be seen that there are bright spots and dark spots on the surface of the photosphere, such bright and dark spots are caused by different temperatures on the surface of photosphere. The dark spots are called "Sunspot", the bright spots are called "Solar Facula". Generally speaking, where there is sunspot, there is Solar Facula. However, the Solar Facular sometimes appears in the area without sunspot.

The Solar Facula

The Solar Facula associated with the sunspot presents fibrous structure, it is 5000~10000km in width and about 50000km in length and it is about 3 times longer than sunspot in life span. The Solar Facula not associated with sunspot presents a circle. It is quite small in size, its diameter is about 2300km and its average life span is about half an hour. The temperature of Solar Facula is higher than the temperature of photosphere. On the other hand, as the Solar Facula is not in a radiation balance, the temperature at the bottom is relatively low and the temperature of upper layer is relatively high, the upper layer of Solar Facula is close to the edge of solar disk, its average temperature is about 100 ℃ higher and its luminance is about 10% brighter than the surrounding area. When the Solar Facula expands outward to the chromosphere, it becomes flocculus. Same as the sunspot, the Solar Facula also has 11-year solar cycle.

4. Granules

There are intense activities in the atmosphere of the photosphere. With aid of telescope, it can be seen that there are many very dense and spotted structure on the surface of the photosphere, like rice grains, they are called "granules". The granule is a solar disk structure on the photosphere, it presents polygonal granule shape. Its diameter stands between 1000~3000km and the temperature is 300-400°C higher than the mean temperature of the photosphere. The granules rise and fall intensely, they are extremely volatile and very short in life span, lasting only 5 to 10 minutes in life span. When the old granules disappear, the new ones will come up, just like the hot bubbles on the boiling rice porridge. The granules presents a cellular texture, the size of a texture is about equal to the size of Texas.

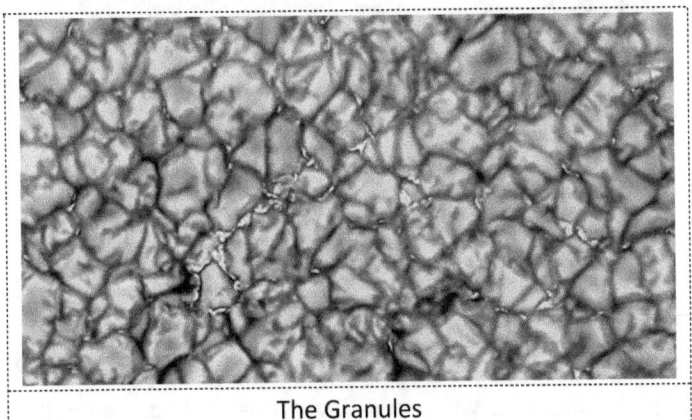

The Granules

It is believed that such granulation is the outcome caused by the intense convection of the gases under the photosphere or is very likely to be the warm gas mass rising from the Convective Zone to the photosphere. They remain unchanged and distributed evenly, and. The discovered super-granule is 30,000 km in diameter and has a life span of 20 hours or so.

5. Solar Wind

The Solar Wind is a plasma flow that exists continuously, coming from the Sun and moves at the speed of 200-800km/s. Different from the earth atmosphere made up of gas molecules, the Solar Wind is made up of elementary particles -- proton and electron one-level smaller than atom. As the effect produced by the flow of such particles is very similar to the effect produced by the air movement, they are called "Solar Wind".

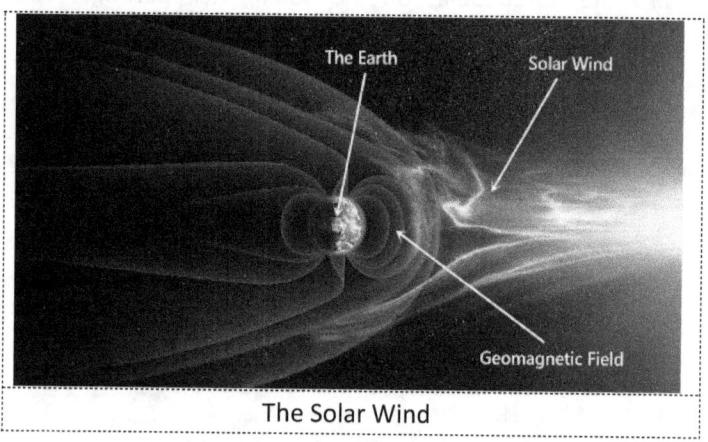

The Solar Wind

Compared with the density of the earth atmosphere, the Solar Wind is very thin and tiny in this respect. Generally, in the interstellar space nearby the Earth, the density of the Solar Wind is several to tens of particles per cubic centimeter, while the density of the earth atmosphere is 268.7 billion molecules per cubic centimeter. Although the Solar Wind is very thin, its might is much far greater than the typhoon on the Earth. The speed of hurricane is above 32.5 meters per second, the speed of solar wind stands at 350~450km/s nearby the Earth, ten thousand times the speed of hurricane. The maximum speed of solar wind can reach more than 800km per second.

The solar wind is a continuous particles stream projected from the corona, the most outward layer of solar atmosphere, to the outer space, such particles is spurted from the coronal hole, and its main compositions are hydrogen particle and helium particle. The solar wind can be divided into two kinds: the first one is the solar wind that radiates out constantly at a low speed and with few particles, it is called "Constant Solar Wind"; the second one is the solar wind that radiates out along with the solar activity at a fast speed and has more particles, it is called "Disturbing Solar Wind". The Disturbing Solar Wind produces great impact on the Earth. When it reaches the Earth, it will trigger very big geomagnetic storm and strong aurora, and produce ionosphere disturbance.

The increase of solar wind and solar activities will severely interfere the radio communication on the Earth and normal working of aerospace equipment, damage the precise electronic equipment, disturb the ground network communication and power control network, and even endanger the astronauts in the spacecraft and space station.

6. Coronal Holes

The coronal holes are distributed in the most areas of the solar surface, particularly in the two polar regions of the Sun. The scientists discovered that there are the closing and opening of magnetic field lines in the coronal hole. If the magnetic field lines suddenly open or close, the solar surface will be covered by wide coronal holes, and the coronal holes distribution in solar surface is far larger than in two polar regions. When the coronal holes come into being, massive scorching plasmas will be produced, the temperature is extremely high, reaching several million Celsius degrees.

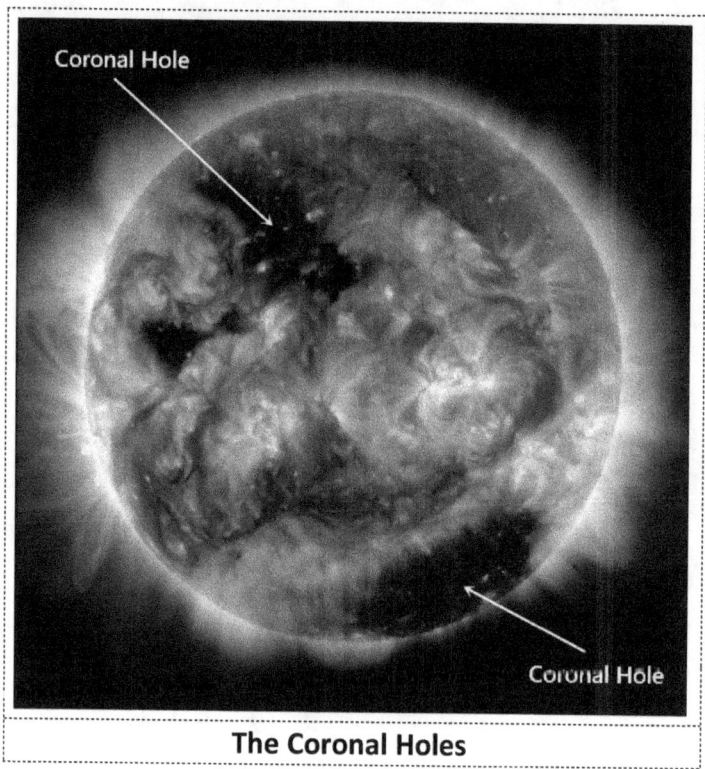

The Coronal Holes

In the areas where the magnetic field lines are open, the spindrift-like structure will appear around the coronal holes. When the coronal holes are distributed in the high latitude areas of the Sun, the solar wind with fast speed will form.

7. Sunlight

Besides the atomic energy, volcano, earthquake and tides intrinsic to the Earth, the sunlight is the primary outside source of the energy received by the Earth. The energy received by one unit area directly exposed to the sunlight is about 1,368W/m^3. After being absorbed by the atmosphere, the energy of the sunlight reaching the surface of the Earth has weakened to 1,000W/m^3 under vertical radiation when the atmosphere is clear.

The Sun transmits the light and heat to the Earth all the time so that the plants on the Earth can make photosynthesis. The leaves of most plants are green because they contain chlorophyll. The chlorophyll uses the energy of the sunlight to synthesize various organisms, this process is called photosynthesis. According to the relevant figures, the green plants on the Earth can produce about 400 million tons of protein, carbohydrate and fat a day, and release nearly 500 million tons of oxygen to the air a day, supplying adequate food and oxygen to human and animals.

Chapter IV The Origin and Evolution of the Sun

In Big Bang, the explosion of black hole causes its inner core and shell substances to make fission reaction in the powerful explosion, the fragment produced in the explosion expands quickly, its volume grows at geometric ratio and expands to the size several billion times the original shape. In the course of fission, the gaseous mass containing massive protium and other substances capable of producing fusion comes into being, it is the nebula. When the substances in the nebula reaches a certain level, the volume and internal pressure of the nebula rise to a specific extent, the nuclear fusion comes up in the nebula, thus forming the embryo of a star. In the long period of time, the star grows up and expands all the time in merging with other stars or absorbing the fragments in the cosmic space in the long travel, and gradually evolves into the sun today.

Nebula (Gaseous Mass) --- The Embryo of A Star

The quick expansion of the fragments is a process of fission in fact. In the fission process, some fragments are preserved in a solid state and they encounter the other solid fragments at all times. In the course of mutual attraction, they make the physical changes or chemical changes and are emerged into a new larger solid object. The new larger solid object keeps on absorbing the smaller solid or liquid substances, its volume and mass increase continuously, and gradually it is capable of capturing and attracting the other objects, in the end, it attracted and captured a little smaller solid object, yet the captured object which has some anti-gravity capability, thus forming the planet and satellite system. The Earth is likely to come into being in this way.

The stars have their own life history from birth and growth to gradual aging and final death although they are different in size, colors and evolution history. The value of stars to life is more than to provide light and heat, the heavy atom that makes up the stars and life substances is created in the explosion of some stars at their last minute. The Sun, as an ordinary star in the universe, also has the same or similar evolution history as other stars. The relevant research findings show that a T Tauri star (TTS) of third-generation first stellar population is formed in the quick collapse of a giant hydrogen molecular cloud about 4.57 billion years ago, it is the Sun. The new star revolves at a nearly circular orbit about 27000 light years away from the Galactic Center.

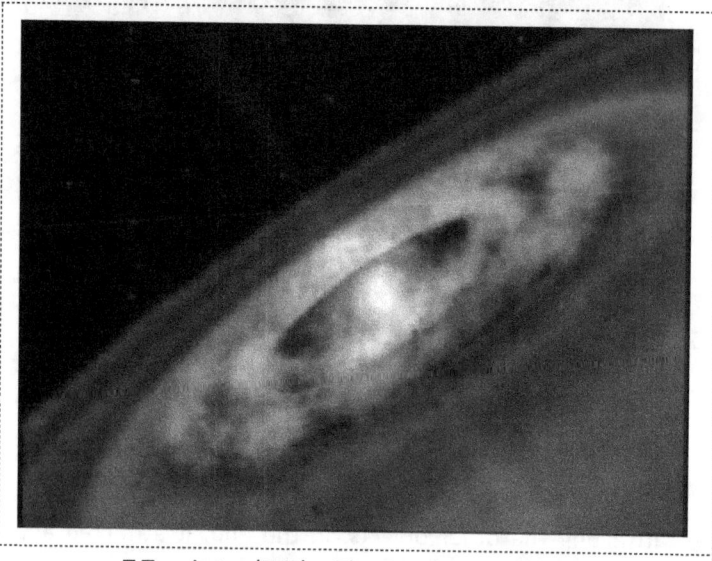

T Tauri star (TTS) – The Sun at Initial Stage

The time when the Sun came into being is measured by two methods: (i) with aid of computer modeling of star evolution and Big Bang nucleosynthesis (BBN), it can be confirmed that the age of the Sun at the main-sequence band is about 4.57 billion years; (ii) with aid of radioactive dating method, it is confirmed that the age of the oldest substance of the Sun is 4.567 billion years. The two outcomes on the age of the Sun are very close or almost same. The Sun now is at middle age in the evolution stage of its main-sequence band, the nuclear fusion at this stage is to convert hydrogen into helium at the core. At the core of the Sun, more than 4 million tons of substances are transformed into energy per second, producing the neutrinos and solar radiation. At this rate, by so far, the Sun has transformed the substances equal to the mass of about 100 earths into energy. The duration of the Sun at the main-sequence band is approximately 10 billion years.

Chapter V The Future of the Sun

As a star of medium mass, the Sun has no sufficient mass to erupt into a supernova and it has the same evolution process and future of other stars of low and medium mass.

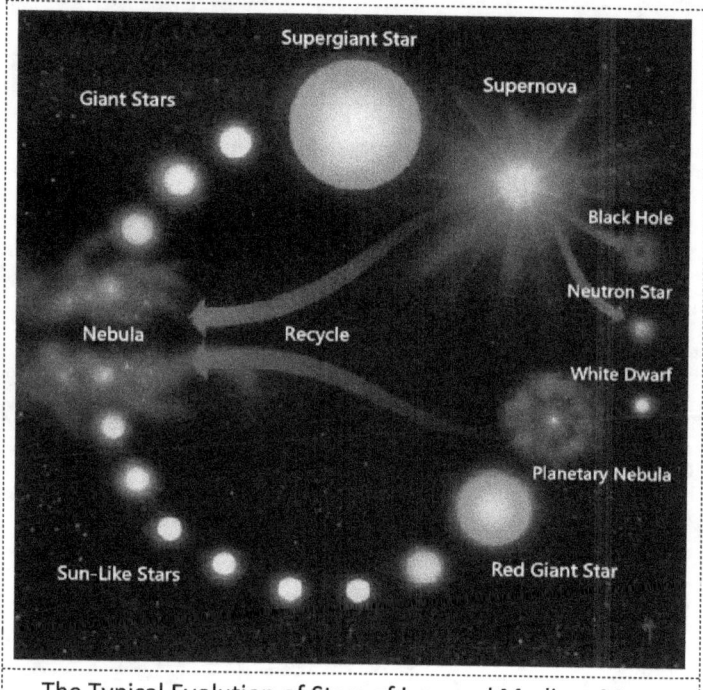

The Typical Evolution of Stars of Low and Medium Mass

About 5 billion years later, the hydrogen in the Sun will become exhaustible, the primary element in the core of the Sun is the Helium, the Sun will become a red giant star or enters into red giant star stage. The Helium core shrinks for resisting the gravity and becomes hot; the hydrogen envelope next to the core accelerates the fusion due to the rise of temperature, the heat produced from the fusion increases continuously and spreads outward, the outer zone of the Sun will expand consequently. When the temperature at the core rises to 100 million k, the helium fusion will start and produce carbon as a result. As the Helium core at this moment is equal to a small white dwarf (electronic degenerate matter), the helium fusion under thermal runaway will trigger the helium flash, the huge energy released therefrom swells the solar core substantially, the electronic degenerate matter is removed, and then the remaining helium at the core will undergo stable fusion. Looking from outside, the Sun will, like a new star, suddenly brighten by 5~10 magnitudes (compared with the luminance when the Sun becomes a red giant star), subsequently, the Sun will shrink greatly in volume and become much more dimmer than the red giant star stage in luminance (but will be more bright than the Sun at middle age), till the carbon at the core gradually accumulates, the Sun enters into the stage of core shrinkage and outer shell expansion, it is the stage of Asymptotic Giant Branch (AGB Stage) of the Sun. After the whole Helium is transformed into the carbon, the fierce thermal pulse will drive away the gases in the outer layer of the Sun, the Sun will lose its shell, forming the planetary nebula. After the shell is lost, the core will collapse, the Sun turns into a white dwarf with the same size as the Earth and a very high density of 10 tons per cm^3, the white dwarf is the last stage of the Sun and other stars. The white dwarf will gradually cool and become dim in the subsequent several billion years, and will disappear in the darkness in the end. It is the typical evolution and future of the stars with low and medium mass, including the Sun.

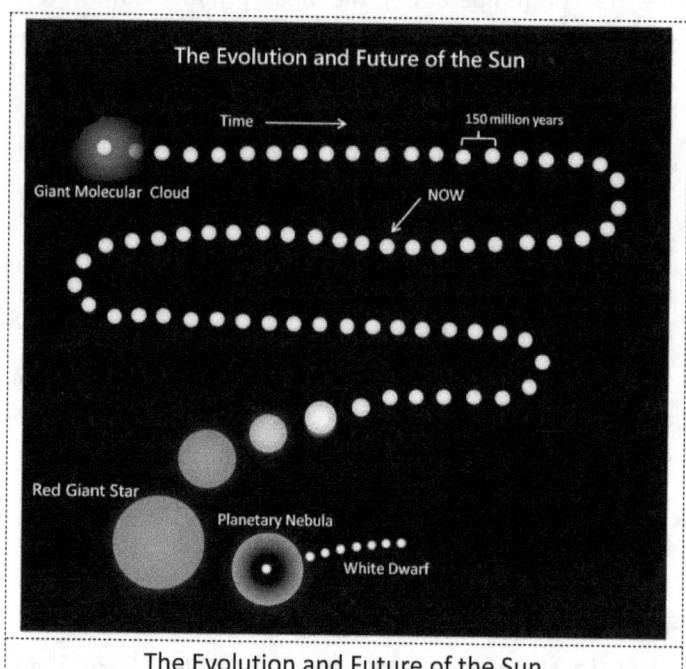

The Evolution and Future of the Sun

Although the Sun will disappear in the end, the fate of the Earth is still uncertain. When the Sun turns into a red giant, its radius will exceeds 2.42 AU (1AU=1.5×10^{11}m) and will be 260 times the radius today, transcending the Earth orbit now. However, when the Sun becomes a star of Asymptotic Giant Branch, it will lose about 30% of its mass due to the action of stellar wind and its gravity to the Earth will decline accordingly, the Earth orbit will move outward. If so, the Earth may be relieved from being absorbed by the Sun. Nevertheless, the new research findings conclude that the Earth is still likely to be absorbed by the Sun due to the tidal action. Even if the Earth can escape such doom, the water on the Earth will be evaporated totally and the atmosphere will dissipate into the outer space.

The Size When the Sun Turns into a Red Giant

The most of hydrogen in the Sun is burning into Helium, the Sun is at the most stable main-sequence star stage. For the Sun as a star of medium mass, its duration at the main-sequence star stage can last 10 billion years. The star will gradually shrink in giving out the light. In the course of shrinking, the density of the core will increase and the pressure will grow, the hydrogen is burned more quickly, the solar surface temperature will rise slowly, the Sun will gradually become brighter. As a matter of fact, since its entry into the main-sequence star stage 4.5 billion years ago, the luminance of the Sun is on a slow increase (at the rate of 10% every one billion years) and increased by 30% by so far, and it will continue to increase in the years to come. The previous luminance of the Sun is relatively dim, which might be the reason why the life emerged on the

Earth one billion years ago. If the solar temperature rises at such rate, the Earth will become too warm in the next one billion years, the water will no longer exist on the Earth in liquid state and all the lives on the Earth will go to extinction.

When the main-sequence star stage of the Sun comes to an end 6.5 billion years later, it is estimated that the luminance of the Sun will be 2.2 times higher than now and the mean temperature of the Earth will be 60°C higher than now, the Earth will be dried up. By then, the Mars may be the planet right for human settlement.

When the Sun is at the main-sequence star stage, the inward shrinking force arising from the gravity of the Sun and the outward force arising from the burning of hydrogen are checked and balanced. However, about 6.5 billion years later, the hydrogen in the Sun will be burned up and only the shell has the hydrogen for burning. As there is no burning area inside the shell, the outward force countering the gravity declines, the core of the Sun will shrink rapidly, the Sun will be increasingly brighter, the shell will expand due to the rise of temperature, the Sun turns into a red giant star or enters into red giant stage. The red giant stage will last several hundred million years, during which the luminance of the Sun will reach 2000 times the luminance now, the temperature around the Jupiter and Saturn will also rise, the icy moon of the Jupiter and the ring as a feature of the Saturn will be vaporized into nothing. In the end, the shell of the Sun will expand to or transcend the Earth orbit now.

On the other hand, the shell of the Sun will release the gas all the time, the mass of the Sun will decrease to 60% of the mass when the Sun is at the main-sequence star stage. Owing to the decline of the gravity of the Sun, the orbits of the planets will move outward. When the mass of the Sun decreases to 60% of its original mass, the distance between the Sun and the planets will increase by 70%, therefore, the Mercury and the Venus are very likely to be absorbed, the Earth will survive as it will be more far away from the Sun due to the decline of the gravity of the Sun to the Earth before the shell of the Sun expands to the Earth orbit now, the Mars and other planets will also survive the red giant stage of the Sun, and they will probably escape the solar system and revolve around another star, which may be a new sun, and a new solar system is likely to be formed.

Chapter VI The Possible New Sun

After the Sun turns into a white dwarf, the planets surviving the red giant stage of the Sun will escape and move about in the cosmic space, the solar system will longer exist. When these planets are attracted and captured a new star, they will revolve around the new star, the new star is another sun and a new solar system may be formed just like now. Who will be the new sun? The astronomical research indicates and astronomers believe that the Jupiter is likely to a new sun after the disappearance of the Sun now. The reasons and facts supporting such opinion are the special characteristics and conditions the Jupiter enjoys.

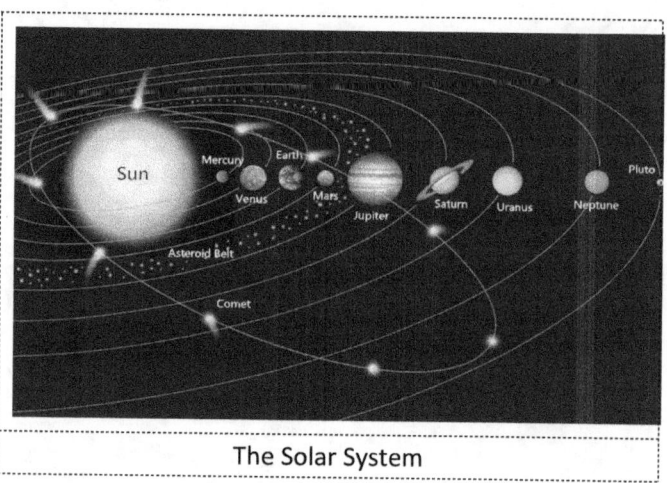

The Solar System

The Jupiter

The Jupiter is the largest planet in the Solar System, its volume is more than 1300 times the volume of the Earth and its mass is also amazingly big, about 2.5 times the total mass of other planets, furthermore, the Jupiter has 61 satellites. Although the Jupiter is very large in size, it is very weak and feeble, its mean density is less than one fourth of the mean density of the Earth and the weight of its matter per cubic centime is only 1.33 grams on average, slightly heavier than the water on the Earth. These facts indicate that the Jupiter is a liquid planet, it has no solid surface just like the continents on the Earth, its surface is the "ocean" of liquid hydrogen. The reason why the hydrogen on the Jupiter exists in liquid state is that the Jupiter is too heavy in weight and too large in volume.

Besides its large size and heavy weight, the Jupiter is a special planet that has some characteristics of a star.

First, the surface temperature of the Jupiter is -148°C, however, by calculating the energy received by the Jupiter from the Sun, its surface temperature should be -168°C, the temperature difference between the two is 20°C, that is, the real surface temperature of the Jupiter is 20°C higher than the surface temperature of it as a planet of solar system;

Second, the research findings show that the luminance of the Sun is on the decline for several thousand years, while the luminance of the Jupiter increases by 2% very year. Based on these two special characteristics of the Jupiter, it can be inferred that there is a heat source inside the Jupiter. The research findings show that the energy released by the Jupiter is the double of the energy received by it from the Sun, the Jupiter can radiate its own energy to warm its surface. In this connection, Jupiter is more than a planet. May scientists believe that the Jupiter is not a planet in strict sense and conclude that it will evolve into a planet in real sense;

Third, The Jupiter is made up of liquid hydrogen and helium and it has the similar atmospheric composition as the Sun has. The scientists point out that the Jupiter is absorbing the massive interstellar gas and dust on the strength of its own huge gravity, although it is only 0.1% of the Sun in respect of volume and mass now, the mass of the Jupiter is on the increase. Once the mass of the Jupiter reaches 80 times the mass now, the substances inside the Jupiter will make thermonuclear reaction, the Jupiter will expand outward and its 61 satellites will be engulfed first. Furthermore, the Jupiter is absorbing the heat energy all the time, its energy is accumulating and growing, the Jupiter will become more warm and bright. When the Sun is closing to its end about 3 billion years later, the Jupiter will turn into a star; when the Sun throws away its shell, the mass of the Jupiter will be the 2.5 times the mass now, the Jupiter has adequate gravity to capture and gradually absorb the shell of the Sun, and attract and capture the other surviving planets, including the Earth, it turns into a "New Sun" (standard star) and replaces the Sun now. By then, the "Jupiter Light" rather than the sunlight will shine the surviving planets of the Sun, and the "Jupiter System" similar to or same as the solar system is likely to form.

As for the question whether the Earth will remain unchanged or same as now or turn into an ordinary planet without living things, it is subject to such conditions: the Jupiter has the same mass as the Sun now; the distance between the Jupiter and the Earth remain unchanged; the light and heat received by the Earth from the Jupiter remain same as now or have no big change.

There is also another possibility that the Earth will turn into a planet like Mars or other planets, and one of the surviving planets or some new planet captured by the Jupiter will turn into the Earth. If so, the origin and evolution of life will restart on the new Earth.

In conclusion, all things are under constant change and evolution, everything, including life, is just a recycle.

---- The End ---

www.ingramcontent.com/pod-product-compliance
Lightning Source LLC
Chambersburg PA
CBHW070841220526
45466CB00002B/848